The Farm Animal Counting Book

A Bedtime Counting Book for Children and Toddlers
Age 2 - 5

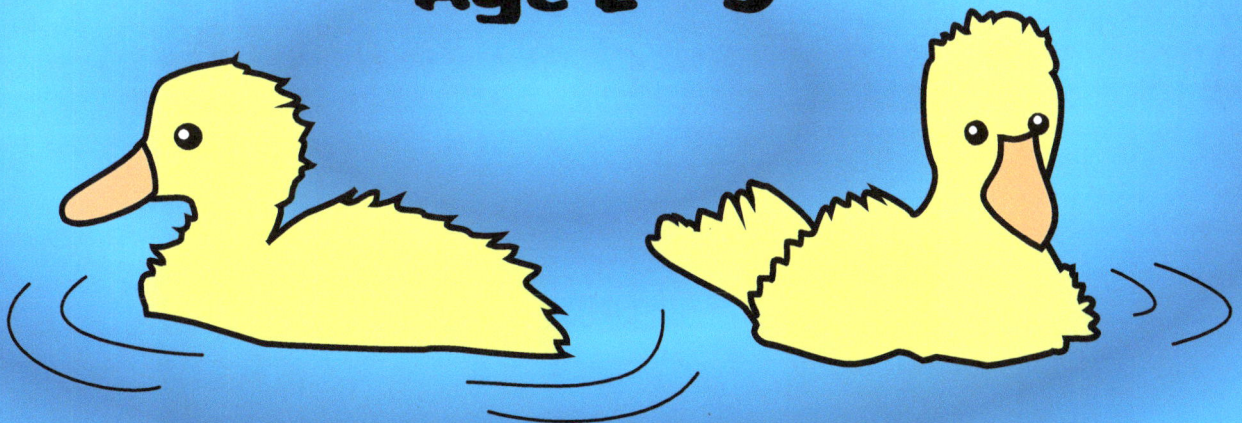

Alina Niemi

Find more great books for kids that mix learning and fun at
http://www.alinaspencil.com

For Charlotte and Wilbur, and all the kids that love them

ISBN: 978-1-937371-07-4

Alina's Pencil Publishing

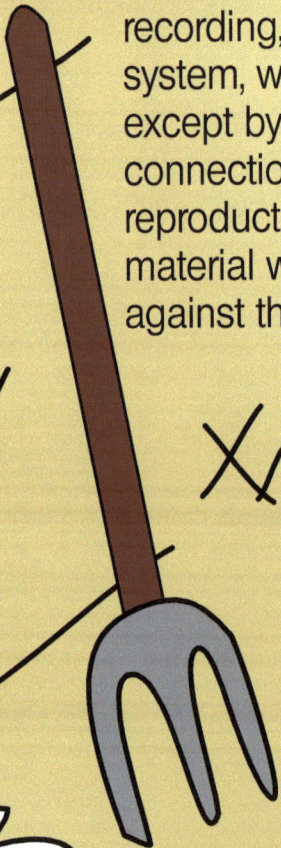

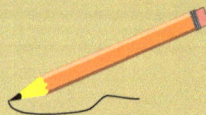

Life on the farm is busy all day
You must do your work before you can play
So while you are here, before this book ends,
Please help me to count our animal friends

1

1 one

1 rooster crowing
to welcome the day

Cock-a Doodle-Doo

Home, Sweet Home

2 two

Two horses munching
with mouths full of hay

4

5

3 three

**Three kittens stretching
and ready to go**

4 four

**Four barn mice hiding
in holes just below**

meow

squeak
squeak

7

5 five

**Five fuzzy ducklings
afloat on a pond**

peep peep

9

6 six

Six grazing cows
in the pasture beyond

moo

11

7 seven

Seven pigs squealing, muddy and plump

weeeee

13

8 eight

**Eight sheep with lambs
who are learning to jump**

maa
maa

9 nine

Nine chickens scratching the dirt in their coop

Aroo?

17

10 ten

**Ten puppies snoring,
asleep in a group**

zzzzzz

As the sun sets,
A long day is done.
I'm happy you came.
I hope you had fun.

It's time now to sleep.
We've come to the end.

Goodbye 'til I see you the next time, my friend.

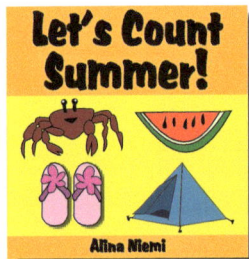

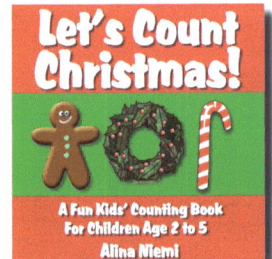

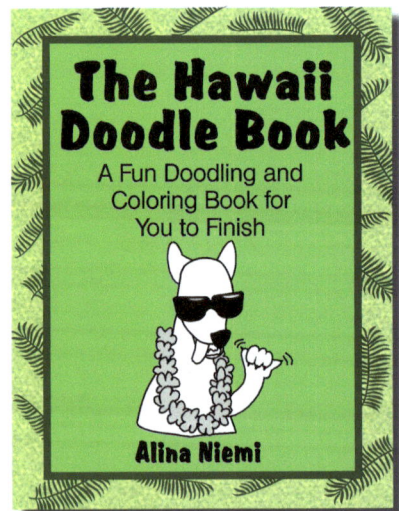

www.ingramcontent.com/pod-product-compliance
Lightning Source LLC
Chambersburg PA
CBHW042113040426
42448CB00002B/252